Sustainable Pathways Toward Hydrogen and Synthetic Fuels 2022

Topical Conference at the 2022 AIChE Annual Meeting

Phoenix, Arizona, USA
13-18 November 2022

ISBN: 978-1-7138-7917-6

Printed from e-media with permission by:

Curran Associates, Inc.
57 Morehouse Lane
Red Hook, NY 12571

Some format issues inherent in the e-media version may also appear in this print version.

Printed with permission by Curran Associates, Inc. (2023)

For permission requests, please contact AIChE
at the address below.

AIChE
120 Wall Street, FL 23
New York, NY 10005-4020

Phone: (800) 242-4363
Fax: (203) 775-5177

www.aiche.org

Additional copies of this publication are available from:

Curran Associates, Inc.
57 Morehouse Lane
Red Hook, NY 12571 USA
Phone: 845-758-0400
Fax: 845-758-2633
Email: curran@proceedings.com
Web: www.proceedings.com

TABLE OF CONTENTS

SUSTAINABLE PATHWAYS TO CLEAN HYDROGEN AND SYNTHETIC FUELS I

SUSTAINABLE PATHWAYS TO CLEAN HYDROGEN AND SYNTHETIC FUELS II

SUSTAINABLE PATHWAYS TO CLEAN HYDROGEN AND SYNTHETIC FUELS III

SUSTAINABLE PATHWAYS TO CLEAN HYDROGEN AND SYNTHETIC FUELS IV

288a Integration of Green Hydrogen Value Chains for Energy-Intensive Industries with Wider Energy Systems: Whole-Systems Modelling and Optimisation

Sheila Samsatli, Nouri John Samsatli

Establishing a thriving low-carbon hydrogen economy is a cornerstone of the UK Government's plan for the country's transition to net zero, as set out in the Prime Minister's Ten Point Plan for a Green Industrial Revolution [1]. The UK Hydrogen Strategy [2] and the UK Net Zero Research and Innovation Framework [3] provide a framework for achieving the UK's target to deliver 5 GW of low-carbon hydrogen production by 2030. Meeting this target requires rapid deployment of low-carbon hydrogen production from a starting point of practically nothing (beyond demonstration projects); current hydrogen production is mainly from natural gas, without carbon capture, and is used as a feedstock for manufacturing chemicals, such as ammonia for fertilisers, and not for energy. There is a need for whole-system value chain analyses to determine the most effective strategies for the rapid and large-scale deployment of low-carbon hydrogen and to understand the systems-integration requirements as well as the implications and impacts on wider energy systems.

Switching energy-intensive industries, such as steel, to low-carbon hydrogen can kickstart a thriving hydrogen market and could be an effective strategy to integrate hydrogen into the wider energy system. This is because these industries have large and reliable energy demands, which if met predominantly from low-carbon hydrogen would encourage many producers into the market. This would facilitate trickle-down of low-carbon hydrogen to other sectors, such as residential heating and mobility. Energy-intensive industries account for 22% of the global CO_2 emissions; steel accounts for 7% and is critical for achieving net zero, as it is used in virtually every technology needed for net zero, such as wind turbines, electric and fuel cell vehicles, solar panels and heat pumps. Further, these sectors are where the largest CO_2 abatement per unit cost can be achieved: e.g., using 1 tonne of green hydrogen to produce steel will abate 26 tonnes of CO_2, whereas only 6 to 7 tonnes of CO_2 would be abated if it were used for mobility, and the abatement is lower for other sectors. Furthermore, there are good electrification options for domestic heating and mobility, e.g. heat pumps and electric vehicles, whereas it is very difficult to electrify completely some industries.

The aim of this study is to determine promising and robust low-carbon hydrogen value chains for hard-to-decarbonise sectors, focusing on the steel industry, and how these value chains can be effectively integrated with those of other sectors (sector coupling). Process models of an integrated steel plant using Direct Reduction of Iron with hydrogen combined with Electric Arc Furnace, which is a net-zero route for producing steel if using green hydrogen, were developed to understand its techno-economic and environmental performance. The results from the process modelling were then integrated into a whole-system value chain optimisation model [4,5] with a detailed representation of the industry, power, buildings and transport sectors, in order to explore how the hydrogen value chains for low-carbon steel and other sectors of the wider energy system could develop in the short- (2020-2025), medium- (2025-2030) and long-term (2030s and beyond). The model determines the most effective planning, design and operation of value chains for multi-sector provision of electricity, mobility, heat, chemicals and other products and services, modelling a wide range of technologies, infrastructures and resources. The model has a high spatial and temporal resolution, which allows representation of local/regional variations in demands and resource availability (e.g. specification of the amount of available/suitable areas for wind, solar and biomass at the 1 km level), detailed transmission and distribution network design (including hydrogen and CO_2 pipelines), planning and investments in technologies (e.g. where, when and how much capacity to invest in), management of storage inventories and operation of all technologies and the entire network at the hourly, daily, seasonal and yearly level. A number of scenarios were modelled and optimised, and in this conference, we will present results that answer the following questions:

- Can sufficient low-carbon hydrogen be produced in the UK for the steel industry?
- What is the optimal mix of green and blue hydrogen to minimise costs and environmental impacts?
- How much offshore, onshore wind and solar capacity, and biomass will be needed?
- How to ramp up demands in low-carbon hydrogen and what are the roles that technologies could play in achieving the levels of production needed to meet the 5 GW target by 2030 and increasing this beyond and to 2050 to meet net-zero commitments?
- How will the hydrogen network and/or carbon capture, utilisation and storage (CCUS) value chains develop and expand?

References:

[1] UK Government, Department for Business, Energy andamp; Industrial Strategy. The Ten Point Plan for a Green Industrial Revolution, November 2020. https://www.gov.uk/government/publications/the-ten-point-plan-for-a-green-industrial-revolution/title

[2] HM Government. UK Hydrogen Strategy, August 2021. https://www.gov.uk/government/publications/uk-hydrogen-strategy

[3] HM Government. UK Net Zero Research and Innovation Framework, October 2021. https://www.gov.uk/government/publications/net-zero-research-and-innovation-framework

[4] Samsatli, S. andamp; Samsatli, N.J., 2019. The role of renewable hydrogen and inter-seasonal storage in decarbonising heat – Comprehensive optimisation of future renewable energy value chains. *Applied Energy*, 233-234, 854-893. DOI: 10.1016/j.apenergy.2018.09.159

[5] Samsatli, S. andamp; Samsatli, N.J., 2018. A multi-objective MILP model for the design and operation of future integrated multi-vector energy networks capturing detailed spatio-temporal dependencies. *Applied Energy*, 220, 893-920. DOI: 10.1016/j.apenergy.2017.09.055

288b Safety and Security Analysis of a Potential Remote Offshore Green Hydrogen Production

Ahmed Elkady, Faisal Khan, Mahmoud El-Halwagi, Moustafa Ali

Green hydrogen is a great potential to decarbonize and support the growth of renewables. Hydrogen production through water electrolysis is a novel hazardous process with limited testing. The operation of such hydrogen facility face extreme operating conditions. Consequently, it's safer to do such operation remotely and with as low as reasonably practical people on board. A great solution to that is to utilize the capabilities of the recent technologies that accompanied the 4th industrial revolution such as digital twins (DTs), Internet of things (IoT), and cyber-physical systems (CPSs). This shall manage the green hydrogen production and process system hazard. However, such relatively new concepts raise new safety and security challenges.

The current work is on the aspect of analyzing cyber security and safety challenges that arise from remote offshore green hydrogen production. It analyzes the literature to investigate opportunities and challenges presented by the application of CPSs in hydrogen production facilities. The analysis observed that technological enablers include:

- Digital twins (DTs): model multiple designs and scenarios, including variables such as weather, to optimize the process, maximize return on investment and minimize risk.
- Internet of things (IoT) and cyber-physical systems (CPSs): monitor process using key performance indicators (KPI) such as energy consumption and production rates which facilitates rabid anomaly detection. It can also help in safety compliance management and prevent violation acts and offer real time alert management systems. Finally, with remote control capabilities, IoT can help in controlling safety levels during normal and emergency conditions

CPSs have characteristics that introduces safety and security challenges. This includes complexity, heterogeneity, interdependency between cyber and physical processes. CPS autonomous reconfiguration decisions and compensatory actions may cause fault masking effects. There are also security-related issues including integrity and availability challenges in the presence of cyber-physical threats with unknown external disturbances. CPS-enabled offshore hydrogen facilities are susceptible to attacks on their physical infrastructure, and cyber-attacks on their data management layer (through sensors, controller, and actuators) as well as against communication layer. This paper provides a comprehensive risk assessment and management that should proactively identify these potential technical problems so that appropriate actions can be taken to reduce or eliminate the probability and/or impact of these problems.

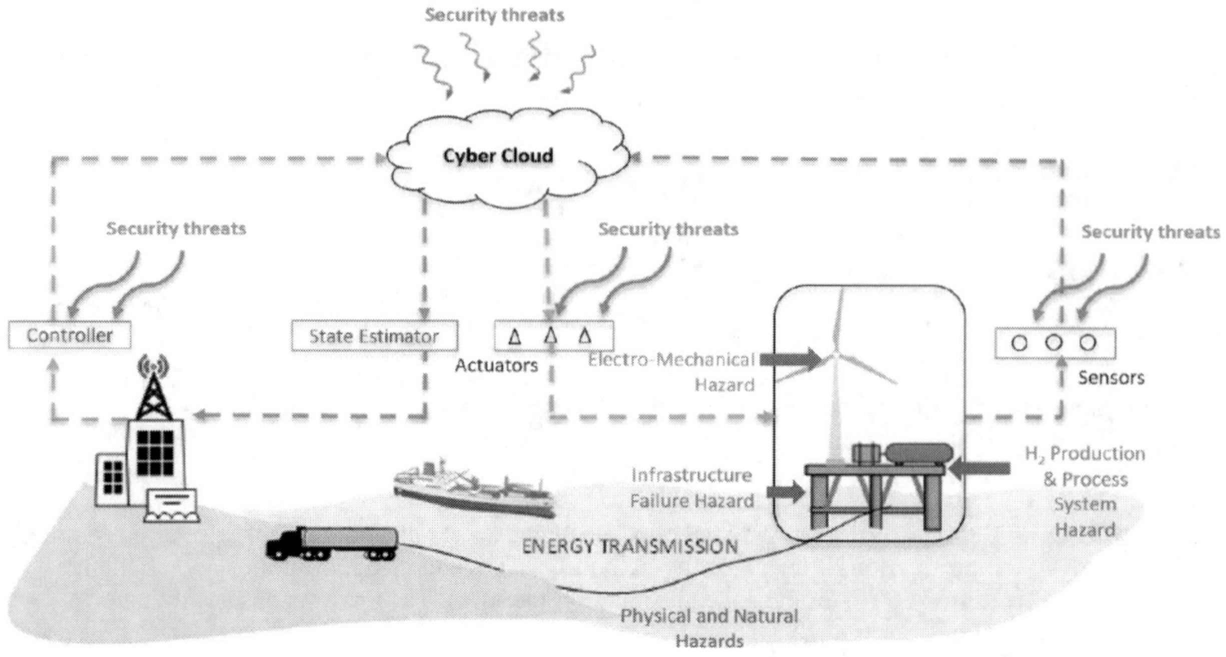
Security threats
Cyber Cloud
Security threats
Security threats
Security threats
Controller
State Estimator
Actuators
Electro-Mechanical Hazard
Sensors
Infrastructure Failure Hazard
H_2 Production & Process System Hazard
ENERGY TRANSMISSION
Physical and Natural Hazards

288c Modeling the Impact of Degradation on the Cost of Low-Temperature Electrolytic Hydrogen Production

Landon Schofield, Benjamin Paren, Doo Hyun Mark Chung, Yang Shao-Horn, Dharik Mallapragada

In recent years, hydrogen (H_2) has gained increasing attention for its potential to act as a low-carbon energy vector and a form of long-term chemical energy storage in a deeply decarbonized energy system. Today, $H_2$is predominantly sourced from fossil fuels, but the increasing demand for energy storage and conversion technologies associated with a future renewables-dominant electric grid necessitates alternative, carbon-free methods of hydrogen production. This has led to significant interest in electricity-driven hydrogen production processes such as low-temperature polymer electrolyte membrane (PEM) electrolysis, with an ambitious scale up of this technology projected over the next decade. Several promising aspects of PEM electrolyzers include their small size compared to traditional alkaline electrolyzers, their ability to operate at high current densities, and their ability to operate with a differential pressure between the anode and cathode, which enables the production of high pressure H_2 product. A number of recent studies have explored how adjusting these characteristics, as well as other operating parameters, affect the long-term performance of PEM electrolyzers. However, in order to optimize the implementation of PEM electrolyzers on a utility scale, it is critical to understand their dynamic operation and the effects of degradation of the stack over time. This presentation will discuss first-principles modeling and techno-economic optimization of PEM electrolysis systems with a specific emphasis on accounting for use-dependent degradation and safety under dynamic operation.

Large-scale electrolyzer projects being contemplated include both "islanded" configurations, where renewable electricity is co-located with the electrolyzer and is the sole source of electricity input, as well as "grid-connected" systems involving using grid electricity as well as on-site or contracted renewable electricity. In both cases, cost-effective operation of the electrolyzer will likely involve operating at much less than 100% capacity utilization to manage fluctuations in available renewable energy [1,2] and grid electricity prices. This partial load operation not only has implications for capital utilization, which has been extensively studied, but also stack lifetime and operating efficiencies that are less well studied. Although there is extensive research on the techno-economic analysis of electrolytic the H_2 supply, for both islanded [3,4] and grid-connected configurations [5], most studies tend to either overlook or abstract out the impact of dynamic operation on stack degradation and lifetime as well as operating efficiency.

In this study we evaluate the factors influencing stack degradation and electrolyzer efficiency at partial loadings and their impact on the levelized cost of H_2 supply via electrolysis in both islanded and grid-connected configurations, using techno-economic modeling. We have developed a first principles electrochemical model for PEM water electrolysis that incorporates an empirical relation for electrolyzer degradation as a function of key operational variables. This model enables characterizing temporal dynamics of production rates but also other key process variables, such as temperature fluctuations and species concentrations at the anode/cathode which are essential for safe operation. The developed model is being incorporated as part of a dynamic optimization model to identify operational routines and design considerations that minimize the levelized cost of hydrogen production considering time-varying electricity prices. Numerical experiments will be shown to demonstrate the value of the proposed model in informing the design and operation of PEM electrolysis systems.

References:

[1] Mallapragada, D.S., Gençer, E., Insinger, P., Keith, D.W., and O'Sullivan, F.M. (2020). Can Industrial-Scale Solar Hydrogen Supplied from Commodity Technologies Be Cost Competitive by 2030? Cell Reports Phys. Sci. 1.

[2] Bødal, E.F., Mallapragada, D., Botterud, A., and Korpås, M. (2020). Decarbonization synergies from joint planning of electricity and hydrogen production: A Texas case study. Int. J. Hydrogen Energy.

[3] Yates, J., Daiyan, R., Patterson, R., Egan, R., Amal, R., Ho-Baille, A., and Chang, N.L. (2020). Techno-economic Analysis of Hydrogen Electrolysis from Off-Grid Stand-Alone Photovoltaics Incorporating Uncertainty Analysis. Cell Reports Phys. Sci. 1, 100209.

[4] Mallapragada, D.S., Gençer, E., Insinger, P., Keith, D.W., and O'Sullivan, F.M. (2020). Can Industrial-Scale Solar Hydrogen Supplied from Commodity Technologies Be Cost Competitive by 2030? Cell Reports Phys. Sci. 1.

[5] Guerra, O.J., Eichman, J., Kurtz, J., and Hodge, B.M. (2019). Cost Competitiveness of Electrolytic Hydrogen. Joule 3, 2425–2443.

288d Solar Thermochemical Water Splitting Using Iron Aluminates

Kent J. Warren, Justin Tran, Alan Weimer

Technologies like solar thermochemical hydrogen (STCH) production, which leverage the oxygen-exchange properties of metal oxides to split water, are attractive because they can harness the entire solar spectrum, whereas photovoltaic-driven alternatives (e.g., water electrolysis) rely only on specific wavelength ranges. Generally, this approach involves alternating between two steps namely, (1) the endothermic removal of oxygen from an oxide and (2) the exothermic oxidation of the reduced oxide with water to produce hydrogen. The endo/exothermic nature of each reaction implies that opposite changes in temperature are required to drive the hydrogen yield per cycle. However, in practice, implementing a temperature swing between redox regimes imposes significant irreversibilities that arise from sensible heating of the solid, thus resulting in inefficient operation unless heat is recovered via ultra-high temperature thermal energy storage (a technology still in its infancy). Therefore, in this work, a scalable prototype reactor – situated within a well-insulated cavity receiver – was designed and operated such that the water-splitting (oxidation) step was intentionally initiated at temperatures where reduction is favorable, thereby avoiding the complications associated with solid-solid heat recuperation. To mitigate the undesired thermodynamic consequence of performing the exothermic oxidation reaction at higher temperatures, iron aluminates, a low-cost class of materials with an exceptional isothermal capacity for hydrogen production, were considered for this prototype demonstration. The stable experimental results, obtained under concentrated radiation from a 45 kW_e high-flux solar simulator, provide compelling motivation for the commercial viability of solar thermochemical routes for hydrogen production.

288e Crystal Features Controlling Oxygen Vacancy Formation in ABO_3 Perovskites

Robert Wexler, Gopalakrishnan Sai Gautam, Ellen B. Stechel, Emily Carter

The control of oxygen vacancy (V_O) formation could unlock significant advances in metal oxide technologies including solar thermochemical fuel production. Despite the critical role played by V_Os in determining the performance of such metal-oxide-based processes, and the research attention they have garnered over the past few decades, an optimally simple, instructive, and efficient model for their quantitative assessment has been elusive. Here, we introduce a compact linear model for the V_O formation energies of ABO_3 perovskites, where A = {Ca, Sr, Ba, Ce, and La} and B = {Ti, V, Cr, Mn, Fe, Co, and Ni}, in six lattice systems (monoclinic, orthorhombic, tetragonal, rhombohedral, hexagonal, and cubic). The model takes as inputs crystal bond dissociation energies, crystal reduction potentials, band gaps, and energies above the convex hull, which can be obtained from theoretical or experimental databases. Additionally, we demonstrate that the model can be simplified, with acceptable losses in accuracy, such that only crystal bond dissociation energies and crystal reduction potentials are needed. Finally, we present our perspectives on how to improve and extend the model, which already provides both accurate and efficient predictions for high-throughput screening and an intuitive and modular phenomenology for renewable energy applications of metal oxide perovskites and beyond.

Wexler, R. B.; Gautam, S. G.; Stechel, E. B.; Carter, E. A. Factors Governing Oxygen Vacancy Formation in Oxide Perovskites. *J. Am. Chem. Soc.* **2021**, *143* (33), 13212-13227. https://doi.org/10.1021/jacs.1c05570

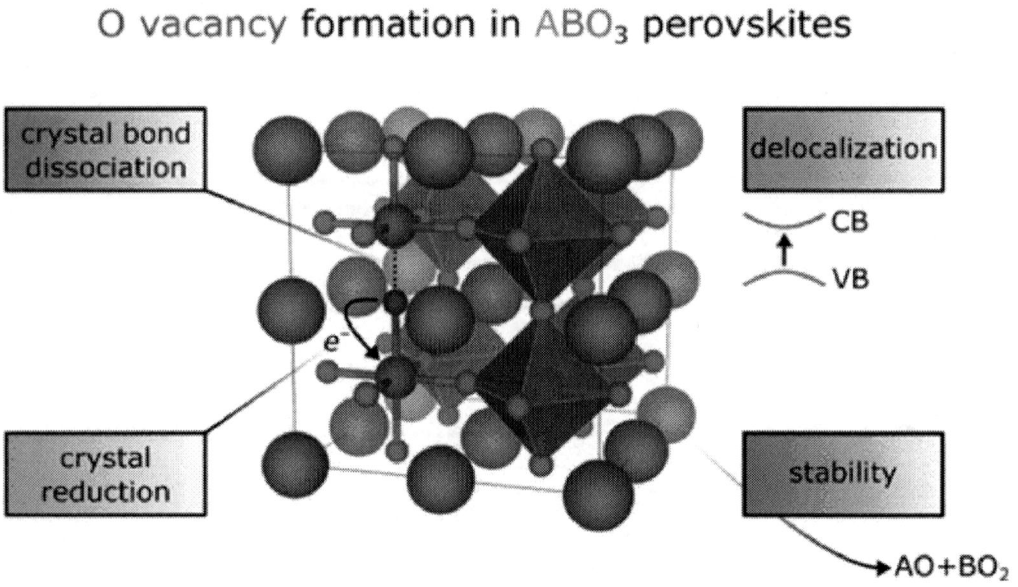

288f Aldehydes Oxidation-Assisted H_2 Production at High Current Densities Promoted By a Cooperative Catalyst

Yue Chen, Yang Hou

Designing highly active electrocatalysts for industry-scale hydrogen production at ampere-level current densities under low overpotentials is still a crucial challenge. Meanwhile, high cost and complex process of technology hinder the large-scale application. Given this, two ways, the decrease of feedstock cost and the production of high value-added byproducts, are potential and practical. Herein, an unique multicomponent nanorods electrocatalyst consisting of minute amount of precious metals and binary alloy nanoparticles anchored on an inorganic metal matrix (Ru-NiMo/NF) for efficient hydrogen evolution reaction (HER) is reported, which delivers a current density of 1000 mA cm^{-2} at ultralow overpotentials of 162 and 179 mV with iR compensation in the alkaline and simulated seawater electrolyte, respectively. Besides, it exhibits a low potential of 1.718 V at 200 mA cm^{-2} for the electrooxidation of 5-hydroxymethylfurfural (HMF), and an optimal 2,5-Furandicarboxylic acid (FDCA) selectivity of over 90%, which significantly improves the efficiency of hydrogen production. This study provides a promising avenue for the safe, efficient and scalable production of high-purity hydrogen.

353a Closing the Carbon Cycle in the Shipping Industry

Margarita Charalambous, Valentina Negri, Valentin Kamm, Gonzalo Guillén-Gosálbez

The shipping industry is highly dependent on fossil-based fuels like heavy fuel oil (HFO), and marine diesel oil (MDO), causing around 2.5% of the global GHG emissions [1]. Hence, keeping its emissions low is key to ensuring the sustainable global trade of goods [2]. Notably, decarbonization efforts set by the International Marine Organization anticipate a GHG-reduction of 70% by 2050 [3]. Hence, adopting alternative fuels i.e., low carbon CO_2-based fuels or zero-carbon fuels, is imperative. In this context, ammonia, hydrogen, and electricity are zero-carbon fuels that hold a promise to halt the use-phase CO_2 emissions from the shipping industry. However, they require new engines and retrofits, which are still under development [4]. Consequently, precedence should be given to CO_2-based fuels.

Motivated by the use of CO_2-based fuels, this work applies circular economy principles in the shipping industry to reduce CO_2 emissions drastically. Notably, since CO_2-based fuels will still emit CO_2 at their use phase, we investigate a new way of reducing the combustion CO_2 emissions by coupling the shipping industry with carbon capture and utilization to close the carbon loop in marine fuels. Following this approach, CO_2 generated on the ship from the engine and steam generation is captured and stored on-board while the ship is moving. Upon arrival at the next port, the CO_2 is unloaded and sent to a chemical plant located near the port. At the chemical plant, CO_2 reacts with renewable hydrogen, generating a CO_2-based fuel which will then be sent to the port to be bunkered by the arriving ships. In this way, the carbon cycle is closed.

A wide range of low-carbon fuels potentially suitable to decarbonize the shipping industry has been identified. The most well-known alternatives to replace HFO suitable for the concept of circular marine fuels are CO_2-based methanol and synthetic natural gas (SNG). Notably, for a CO_2-based fuel to be appropriate for usage in ships, it should contain a high specific energy, low production cost, and should be easily scalable. At the same time, available bunkering infrastructure in most ports will be necessary. The analysis considers three scenarios; CO_2-based methanol for propulsion and steam generation, CO_2-based methanol for propulsion and natural gas for steam generation, and synthetic natural gas for propulsion and steam generation.

The carbon capture technology of choice is post-combustion capture with monoethanolamine (MEA) as the amine-based solvent. For all the scenarios, the process flowsheets were developed in Aspen Hysys v.11. For the carbon capture part, the composition and flow rate of the engine exhaust were calculated using the lower heating value of each fuel and considering a 23000 kW engine. The two streams of flue gases, one coming from the engine exhaust and one from the furnace, are mixed, cooled, pumped, and fed to the post-combustion capture section, which captures CO_2. The fuel demand for the furnace were calculated from the steam requirements in the stripper column. Lastly, the CO_2 is liquefied and stored on-board.

For the liquefaction of CO_2, an ammonia refrigeration cycle is included for the methanol-fueled ship. In contrast, CO_2 is cooled down in the evaporator that heats up SNG for the natural gas fueled ship, while for additional cooling, the ammonia refrigeration cycle is used. For the fuel production plants, methanol production was based on González-Garay et al. [5], and synthetic natural gas on Chauvy et al. [6]. The fuel production of the chemical plants was based on the marine fuel demand of Hamburg's' port.

An economic and environmental assessment was performed as well as a technical feasibility analysis for all the scenarios. For the technical feasibility analysis, the weight and space taken by the equipment added on-board the ship for a week trip were considered and compared to the actual cargo weight (DWT – deadweight tonnage) and space (TEU – twenty-foot equivalent units). In order for these scenarios to be feasible and profitable, the added equipment should not take up most of the space of the cargo, hindering their purpose. The percentage that added equipment takes is up to 5% of the cargo. Regarding the economic assessment, the total cost of each scenario includes capture on-board and fuel production, covering capital and operating expenditures.

The environmental assessments were implemented in SimaPro v9.2.0.2 software [7], following the ISO 14040/14044 standards. The inventories integrate data of the foreground and background systems. For the background system, data available in Ecoinvent v3.5 were used [8]. The mass and energy flows in the foreground system were taken from process simulations complemented with literature data. For the production of the steam generation fuel used in the furnace, CO_2 from direct air capture was modelled according to Keith et al. [9]. The hydrogen used in the chemical plants is produced from polymer electrolyte membrane electrolysis powered by offshore wind electricity [10]. The inventory of the ship construction was also taken from Ecoinvent v.3.5. The goal of the LCA was to compare CO_2-based methanol and SNG as circular marine fuels with the business-as-usual (BAU) based on IPCC 2013 GWP 100a impact assessment method. The scope of the analysis is cradle-to-cradle, and the functional unit (FU) is 1 tonne ·kilometer.

Our results show that the proposed concept can be technically and environmentally feasible, whereas much more economical compared to other proposed decarbonisation scenarios. More precisely, concerning technically feasible, we find that additional equipment will only take up a small percentage of the actual cargo transported. Moreover, regarding the environmental results, CO_2-based methanol for propulsion and steam generation can generate the most emission reductions. Lastly, considering the costs, we find that the cost of capture on-board is minimal in comparison to the production of the fuel, where the economic feasibility depends strongly on the cost of electrolytic hydrogen. Our work opens up a new pathway for tackling the shipping industry's emissions by coupling carbon capture offshore with utilization onshore.

Acknowledgments: This work was created as part of the NCCR Catalysis, a National Centre of Competence in Research funded by the Swiss National Science Foundation.

References

[1] IMO *et al.*, 'Third IMO Greenhouse Gas Study 2014,' *Int. Marit. Organ.*, p. 327, 2014, doi: 10.1007/s10584-013-0912-3.

[2] ITF, *Organisation for Economic Co-operation and Development (OECD) - The ITF Transport Outlook 2021*. 2021.

[3] 'Initial IMO GHG Strategy.' https://www.imo.org/en/MediaCentre/HotTopics/Pages/Reducing-greenhouse-gas-emissions-from-ships.aspx (accessed Apr. 05, 2022).

[4] N. Gray, S. McDonagh, R. O'Shea, B. Smyth, and J. D. Murphy, 'Decarbonising Ships, Planes and Trucks: An Analysis of Suitable Low-Carbon Fuels for the Maritime, Aviation and Haulage Sectors,' *Adv. Appl. Energy*, vol. 1, p. 100008, Jan. 2021, doi: 10.1016/j.adapen.2021.100008.

[5] A. González-Garay, M. S. Frei, A. Al-Qahtani, C. Mondelli, G. Guillén-Gosálbez, and J. Pérez-Ramírez, "Plant-to-planet analysis of CO_2-based methanol processes," *Energy Environ. Sci.*, vol. 12, no. 12, pp. 3425–3436, Dec. 2019, doi: 10.1039/c9ee01673b.

[6] R. Chauvy, L. Dubois, P. Lybaert, D. Thomas, and G. De Weireld, 'Production of synthetic natural gas from industrial carbon dioxide,' *Appl. Energy*, vol. 260, p. 114249, Feb. 2020, doi: 10.1016/J.APENERGY.2019.114249.

[7] M. Goedkoop, M. Oele, J. Leijting, T. Ponsioen, and E. Meijer, "Introduction to LCA with SimaPro Colophon," *Introd. to LCA with SimaPro*, no. November, 2016.

[8] G. Wernet, C. Bauer, B. Steubing, J. Reinhard, E. Moreno-Ruiz, and B. Weidema, 'The ecoinvent database version 3 (part I): overview and methodology,' *Int. J. Life Cycle Assess.*, vol. 21, no. 9, pp. 1218–1230, 2016, doi: 10.1007/s11367-016-1087-8.

[9] D. W. Keith, G. Holmes, D. St. Angelo, and K. Heidel, 'A Process for Capturing CO_2 from the atmosphere,' *Joule*, vol. 2, no. 8, pp. 1573–1594, Aug. 2018, doi: 10.1016/j.joule.2018.05.006.

[10] K. Bareiß, C. de la Rua, M. Möckl, and T. Hamacher, 'Life cycle assessment of hydrogen from proton exchange membrane water electrolysis in future energy systems,' *Appl. Energy*, vol. 237, pp. 862–872, Mar. 2019, doi: 10.1016/j.apenergy.2019.01.001.

353b Economic and Environmental Assessment of Fischer-Tropsch Electro-Diesel from Captured CO_2

Juan D. Medrano-García, Margarita Charalambous, Gonzalo Guillén-Gosálbez

Abstract

The transportation sector is responsible for 8.19 Gt of CO_2 emitted in 2019 due to its dependence on fossil fuels and particularly diesel [1]. Under current policies, there is a strong motivation to replace fossil diesel with more sustainable alternatives. Nevertheless, the good properties of diesel and the current compression ignition engines make the transition particularly challenging. As a substitute of fossil diesel, Fisher-Tropsch electro-diesel (FT e-diesel) is thought to be one of the best alternatives with the highest "drop-in" quality, as it shares similar features with its fossil analog [2]. Notably, FT e-diesel could be integrated into the existing infrastructure with the potential to reduce pollutant emissions [2].

Many previous works on FT e-diesel focused on carbon emissions while disregarding impacts beyond global warming. However, burden-shifting has been observed in many emerging technologies when attempting to mitigate climate change; that is, improvements in some environmental categories inevitably worsen others. Notably, carbon capture and utilization [3], low carbon power mixes [4] and biomass synthetic fuel synthesis [5], to mention a few, have been shown to lead to burden-shifting. This narrow scope of current assessments is also found in FT e-diesel studies, which in view of the former examples, raises the question of this collateral damage being present in the production of the synthetic fuel.

In this work, we assess the economic and environmental performance of FT e-diesel to shed light on the potential occurrence of burden-shifting. We consider six different scenarios resulting from the combination of two CO_2 capture technologies (post combustion capture at a coal power plant, COAL, and direct air capture, DAC) and three electricity sources (wind onshore, solar photovoltaic and nuclear). The material and energy flows in the foreground system were extracted from a simulation in Aspen HYSYS v11, and combined with life cycle inventory data of the background system retrieved from an environmental database. The total annualized cost was computed following well-stablished correlations [6], while the life cycle assessment (LCA) was applied following the ISO 14040 standards [7], using SimaPro v9.2 [8] and Ecoinvent v3.8 [9] to carry out the calculations.

The results show that Fischer-Tropsch electro-diesel from captured CO_2 is economically unappealing (400 – 1000 % increase in cost relative to fossil diesel), with DAC CO_2 being more expensive than the COAL scenario, and solar energy-powered H_2 more costly than nuclear and wind. The inclusion of externalities, that is, the monetization of the ReCIPe 2016 endpoint impacts [10], helps close the gap between COAL and DAC, although regardless of the scenario, fossil diesel still holds the lowest cost.

Regarding the LCA analysis, the global warming impact of FT e-diesel, especially in the DAC and nuclear scenario, decreases drastically with respect to fossil diesel (- 3,87 kg CO_2-eq/kg vs. 0.57 kg CO_2-eq/kg, respectively). The wind-DAC scenario also reduces the impact (- 1.94 kg CO_2-eq/kg), while solar-DAC and all the COAL-related scenarios perform worse than fossil diesel. However, although resource scarcity follows the decreasing trend in global warming impact, both human health and ecosystems quality worsen their performance relative to fossil diesel.

Overall, our results, highlighting the occurrence of burden-shifting in FT e-diesel from captured CO_2, reinforce the need to embrace a wide range of impacts in the environmental assessments of fuels to avoid myopic solutions mitigating greenhouse gas emissions at the expense of exacerbating other impacts.

References

[1] European Union. L 282/4. Official Journal of the European Union 19.10.2016. Paris agreement 2016:1–15.

[2] Schemme S, Samsun RC, Peters R, Stolten D. Power-to-fuel as a key to sustainable transport systems – An analysis of diesel fuels produced from CO2 and renewable electricity. Fuel 2017;205:198–221. https://doi.org/10.1016/j.fuel.2017.05.061.

[3] Algunaibet IM, Guillén-Gosálbez G. Life cycle burden-shifting in energy systems designed to minimize greenhouse gas emissions: Novel analytical method and application to the United States. J Clean Prod 2019;229:886–901. https://doi.org/10.1016/J.JCLEPRO.2019.04.276.

[4] Galán-Martín Á, Tulus V, Díaz I, Pozo C, Pérez-Ramírez J, Guillén-Gosálbez G. Sustainability footprints of a renewable carbon transition for the petrochemical sector within planetary boundaries. One Earth 2021;4:565–83. https://doi.org/10.1016/j.oneear.2021.04.001.

[5] Cuéllar-Franca R, García-Gutiérrez P, Dimitriou I, Elder RH, Allen RWK, Azapagic A. Utilising carbon dioxide for transport fuels: The economic and environmental sustainability of different Fischer-Tropsch process designs. Appl Energy 2019;253:113560. https://doi.org/10.1016/j.apenergy.2019.113560.

[6] Turton R, Bailie RC, Whiting WB, Shaeiwitz JA, Bhattacharyya D. Analysis, Synthesis and Design of Chemical Processes. Fourth. Prentice Hall; 2012.

[7] ISO - ISO 14040:2006 - Environmental management — Life cycle assessment — Principles and framework n.d.

[8] Goedkoop M, De Schryver A, Oele M, Durksz S, de Roest D. Introduction to LCA with SimaPro 7. PRé Consult Netherlands 2008.

[9] Wernet G, Bauer C, Steubing B, Reinhard J, Moreno-Ruiz, E., Weidema B. The ecoinvent database version 3 (part I): overview and methodology. Int J Life Cycle Assessment, [Online] 21(9) 2016:1218–1230. http://link.springer.com/10.1007/s11367-016-1087-8 (accessed January 24, 2020).

[10] Weidema BP. Comparing Three Life Cycle Impact Assessment Methods from an Endpoint Perspective. J Ind Ecol 2015;19:20–6. https://doi.org/10.1111/jiec.12162.

353c Modeling of an Integrated Fischer-Tropsch Process with Energy Supplied By a Nuclear Power Plant

Guiyan Zang, Pingping Sun, Hernan E. Delgado, Vincenzo Cappello, Clarence Ng, Amgad Elgowainy

The Fischer-Tropsch (FT) process is mature process developed as an alternative means to produce liquid hydrocarbon fuels when petroleum crude oil is short of supply. FT fuels can be low-carbon fuels if synthesized from carbon dioxide (CO_2) and green hydrogen as feedstocks. The green hydrogen can be produced via electrolysis of water using low-carbon or zero-carbon electricity, while the CO_2 can be captured from various waste streams (e.g., from industrial facilities or fossil power generators).

Our work [1] modeled the synfuel production process (using Aspen Plus) to produce low carbon FT diesel, jet, and naphtha, evaluated the process economics, and analyzed its environmental impact. Our study considered nuclear energy as the electricity source due to its ability to provide continuous supply of electricity with near-zero carbon emissions and its ability to produce hydrogen with high efficiency via high temperature electrolysis, providing both electricity and steam. Our work assumed CO_2 is collected from industrial plants waste streams with high purity CO_2, which is subsequently transported to the FT plant by pipeline [2]. The CO_2 and H_2 are converted into syngas via a reverse water gas shift (RWGS) reaction, which undergoes FT reactions to produce hydrocarbons that are separated into different fuel pools.

In addition to the process design, a life cycle assessment (LCA) and techno-economic analysis (TEA) were conducted using Argonne GREET and the H2A model to assess the environmental impact and the economic feasibility of the FT process. Our study shows the synfuel production cost is dominated by the H_2 feedstock cost.

A total of three different process scales—sourcing 100, 450, and 1000 MWe from the nuclear power plant—were modeled to assess the impact of scale on economics. The impact of using nuclear energy, and its integration with the FT production process are analyzed, particularly the impact of hydrogen price, and the key role of green hydrogen in reducing liquid FT hydrocarbons greenhouse gas emissions.

References

[1] Zang G, Sun P, Elgowainy A. The Modeling of Synfuel Production Process: ASPEN Model of FT production with electricity demand provided at LWR scale. 2021.

[2] Zang G, Sun P, Yoo E, Elgowainy A, Bafana A, Lee U, et al. Synthetic Methanol/Fischer–Tropsch Fuel Production Capacity, Cost, and Carbon Intensity Utilizing CO2 from Industrial and Power Plants in the United States. Environ Sci Technol 2021;55:7595–604. https://doi.org/10.1021/acs.est.0c08674.

353d A Technoeconomic Assessment of Liquid Green Hydrogen Carrier Supply Chains for Delivering Sustainable Fuel

Bradie S. Crandall, Todd Brix, Robert Weber, Feng Jiao

Sustainably produced hydrogen (green hydrogen) can play a key role in the decarbonization of society, both as a chemical feedstock and as an energy carrier. The rapidly declining costs of renewable electricity generation permit affordable green hydrogen production. However, the storage and delivery costs of gaseous hydrogen pose significant barriers to its economic feasibility. These limitations may be overcome with the aid of one-way liquid green hydrogen carriers, such as ammonia and methanol, which have been frequently studied, as well as formic acid, which has received less attention. Formic acid can be produced sustainably via the electrochemical CO_2 reduction reaction with relatively high selectivity (>80% Faradaic efficiency). Ammonia and methanol both lack high performing direct electrochemical routes but can be produced sustainably by coupling green hydrogen with CO_2 hydrogenation and Haber-Bosch processes, respectively. Techno-economic assessment serves as a useful tool to evaluate each of these liquid green hydrogen carriers. To assess the economics of each carrier, a case study was constructed to deliver green hydrogen as a distributed transportation fuel. This assessment revealed the potential for formic acid produced via electrochemical CO_2 reduction to serve as an affordable means to deliver hydrogen fuel to consumers. This techno-economic assessment can also be deployed to inform researchers how to best accelerate the commercialization of green formic acid. A pathway of modest advances in green formic acid production technology was also crafted to illustrate the economic feasibility of green formic acid as a hydrogen carrier at a large scale (50,000 kg H2/day). Additional investigation also revealed that formic acid is the only green hydrogen carrier that is cheaper to produce than its fossil-based counterpart and that it also offers the safest logistics of the studied hydrogen carriers.

353e Thermal Hydrolysis of Solid-State Sodium Borohydride

Geo-Jong Kim, Hyun Tae Hwang

Hydrogen from sodium borohydride ($NaBH_4$, SBH) can be generated by thermolysis or hydrolysis. Unfortunately, thermolysis of SBH requires high temperatures (>300 °C), while hydrolysis is conducted in the presence of catalyst and self-hydrolysis can cause safety concerns due to unexpected generation of hydrogen. In order for hydrolysis of SBH to be practical, efficient and safe methods to produce hydrogen are required. Here, we propose a new process, thermal hydrolysis of SBH, to generate hydrogen via hydrolysis in the solid phase. In this new process, SBH generates hydrogen by hydrolysis with water produced by thermal decomposition of boric acid (BA). Self-hydrolysis causing safety issues no longer occurs because the SBH-BA mixture in the solid phase is stable at below 80 °C. When heating the mixture, a rapid hydrogen evolution was observed at ~120 °C and a maximum H_2 equivalent of 3.9 was achieved at 150 °C for SBH:BA = 1:4. BA is dehydrated to form metaboric acid with a release of 1 mol H_2O at 150 °C and an additional 0.5 mol water at 200 °C. With recent progress in regeneration of SBH spent fuels, this newly proposed method is promising for hydrogen storage in proton exchange membrane fuel cell applications.

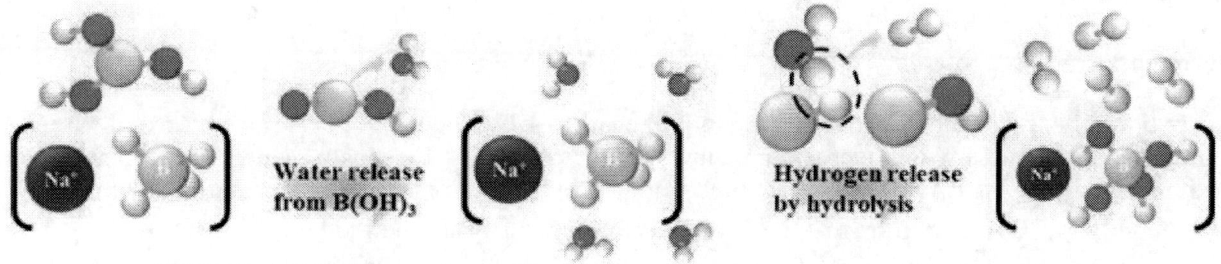

$$B(OH_3) \rightarrow BHO_2 + H_2O \ (150\ ^{\circ}C) \qquad 4NaBH_4 + 4H_2O \rightarrow 4NaB(OH)_4 + 4H_2$$

$$BHO_2 \rightarrow \frac{1}{2}B_2O_3 + \frac{1}{2}H_2O \ (200\ ^{\circ}C)$$

353f Performance of Cyclic Hydrogenation and Dehydrogenation of Benzyltoluene As Liquid Organic Hydrogen Carrier (LOHC)

Timo Rüde, Patrick Preuster, Moritz Wolf, Peter Wasserscheid

Introduction

Transformation of the global energy system requires green hydrogen technologies since they enable energy storage in large quantities and over long periods of time as well as global energy distribution. For such large-scale hydrogen storage and transportation, many technologies have been discussed. The LOHC technology offers a unique selling point, as the hydrogen is reversibly chemically bound in a liquid hydrocarbon carrier by catalytic hydrogenation and dehydrogenation reactions. These hydrocarbons enable utilisation of existing infrastructure for liquid fuels for the hydrogen logistics.[1,2] To date, two hydrocarbon-based heteroatom-free LOHC systems have been intensively studied in the literature, namely toluene/ methylcyclohexane[3,4] and dibenzyltoluene/perhydro dibenzyltoluene.[5-7] Despite the fact, that the benzyltoluene/perhydro benzyltoluene system was already mentioned in 2014,[8] this alternative system has been studied far less. Only recently, two studies examined the H0-BT/H12-BT LOHC system in greater detail.[9,10] It was concluded that H0-BT/H12-BT is particularly interesting as it combines fast production of high purity hydrogen, which is typically associated to dibenzyltoluene-based LOHC systems, with the technically relevant low viscosity and fast pore diffusion characteristics during dehydrogenation, which is typically associated with methylcyclohexane. Motivated thereby, we conduct a more detailed study of the H0-BT/H12-BT LOHC system with analysis of the rate of hydrogen storage and hydrogen release while assessing the stability of this LOHC system in repeated storage cycles.

Experimental

Semi continuous experiments for studying the hydrogenation/dehydrogenation cycles were carried out in a stainless-steel high pressure/high temperature autoclave (Type 4575A, Parr Instruments). The general procedure for performing the cycles was adopted from a previous study focussing on catalyst testing and LOHC stability.[5,11] Hydrogenation reactions were performed in a dead-end mode and a pressure of typically 30 bar_{abs} was maintained in the reactor by a pressure regulator. The dehydrogenation reactions were initiated by stopping hydrogen supply and opening the reactor to a back pressure regulator. For dehydrogenations, the overall pressure was set to ensure a hydrogen partial pressure of 1.6 bar_{H2}. The temperature was controlled by an electric heating mantle and an immersed cooling coil. The utilized LOHC of technical quality was applied without further preconditioning or purification and the finely ground EleMax-102D (Cl#102, 0.3 wt.-% Pt/Al_2O_3, Clariant) served as catalyst.

The degree of hydrogenation (DoH) and the by-product formation were determined by liquid sample analysis by means of gas chromatography (GC). The GC was equipped with mass-spectrometry (MS) and flame-ionisation (FID) detectors to identify and quantify the LOHC species. The DoH of the sample is calculated from GC peak areas using equation (1).

Results and Discussion

Hydrogenation of H0-BT and dehydrogenation of H12-BT in repeated cycles

Here, we demonstrate up to seven repeated storage cycles at 290 °C. Hydrogenation was completed after 4 h with a mean DoH of (99.6 ± 0.2)%. In the subsequent dehydrogenation sequences over 20 h, DoHs of 26.2% were reached as consequence of the relatively elevated hydrogen partial pressure. The prolonged 5th dehydrogenation resulted in a DoH of 17.9%, which represents the chemical equilibrium under the applied conditions.[11] In another experimental series, the H0-BT hydrogenations were repeatedly demonstrated to be 260% more productive than the H0-DBT hydrogenations. The H12-BT dehydrogenations were 30% more productive than H18-DBT dehydrogenations, with the latter showing a decreased hydrogen storage capacity over the cycles, whereas the H12-BT dehydrogenation sequences showed remarkably stable productivities up to a total operation time of 216 h.[11]

Comparing these results from Figure 1 with literature reports on storage cycles with H0-DBT/H18-DBT under identical conditions is quite instructive. Jorschick *et al.* reported significant catalyst deactivation during long-term dehydrogenation.[12] In contrast, the herein reported storage cycles with H0-BT/H12-BT, the catalyst activity and LOHC stability appeared unaltered. Even the long-term 5th dehydrogenation over 70 h reaching equilibrium conversion did not impair the catalyst, while yielding the same DoH (25.6%) in the 6th as in the 4th dehydrogenation.

Monitoring of fluorene formation during extended storage cycles

Exposure of the hydrogenation/dehydrogenation catalyst to LOHC systems at low hydrogen partial pressures in combination with elevated temperatures may result in catalyst deactivation and high-boiling by-product formation.[12] In our experiments, four high-boiling regioisomeric methylfluorenes were the only significant by-products detected by means of GC-MS and GC-FID. Expectedly, their concentration increased with every storage cycle. However, fluorene derivatives themselves have been considered as promising LOHC compounds in the literature reflecting their ability to store/release six hydrogen molecules resulting in a comparable hydrogen storage capacity as the H0-BT/H12-BT LOHC system.[13] Unfortunately, fluorene derivatives can also trigger the formation of even larger condensation products and finally coke. Therefore, we investigated the methylfluorene formation in four storage cycles at 290 °C and 1.0 bar_{H2} during dehydrogenations. By linear regression, an increase of fluorene derivatives with each dehydrogenation of about 0.32%-points can be estimated (Figure 2). The experiments strongly indicate that BT forms less by-products than DBT.[11] Aside from this quantitative analysis, the liquid samples of BT experiments (such as those shown in Figure 1) hardly displayed any visible coloration with increasing operation time. This can be seen as strong indication for the formation of only trace amounts of heavy-boiling by-products, which are distinguishable by a yellow or brownish colouring.

Conclusions

Inspired by the low viscosity of H12-BT, when compared to H18-DBT, we have collected additional technically relevant data to assess the H0-BT/H12-BT system for technical use in the frame of the LOHC technology. In detail, we have studied the stabilities of the LOHC compounds and the performance of a technical Pt-based hydrogenation/dehydrogenation catalyst over repeated storage cycles. We found that the hydrogen storage cycles under benign conditions (290 °C with 30 bar hydrogen pressure in loading and 1.0 – 1.6 bar_{H2} hydrogen pressure in release sequences) showed an excellent reproducibility. Seven storage cycles with one long-term dehydrogenation over several days neither showed signs for degradation of the catalyst nor of the LOHC compounds. We anticipate that the H0-BT/H12-BT LOHC system may represent the preferred hydrocarbon-based and heteroatom-free LOHC system for further technical application with clear advantages over the more traditional toluene/methylcyclohexane and H0-DBT/H18-DBT LOHC systems.

References

1. Niermann, M.; Drünert, S.; Kaltschmitt, M.; Bonhoff, K., Liquid organic hydrogen carriers (LOHCs) – techno-economic analysis of LOHCs in a defined process chain. *Energy andamp; Environmental Science* **2019,** *12* (1), 290-307.

2. Niermann, M.; Timmerberg, S.; Drünert, S.; Kaltschmitt, M., Liquid Organic Hydrogen Carriers and alternatives for international transport of renewable hydrogen. *Renewable and Sustainable Energy Reviews* **2021,** *135*, 110171.

3. Ichikawa, M., Organic liquid carriers for hydrogen storage. In *Solid-State Hydrogen Storage*, Walker,

G., Ed. Woodhead Publishing: Sawston, Cambridge, 2008; Vol. 1, pp 500-532.

4. Okada, Y.; Shimura, M. In *Development of large-scale H2 storage and transportation technology with Liquid Organic Hydrogen Carrier (LOHC)*, Technical paper at Joint GCC-JAPAN Environment Symposia, Japan, Japan, 2013.

5. Jorschick, H.; Preuster, P.; Dürr, S.; Seidel, A.; Müller, K.; Bösmann, A.; Wasserscheid, P., Hydrogen storage using a hot pressure swing reactor. *Energy andamp; Environmental Science* **2017,** *10* (7), 1652-1659.

6. Modisha, P.; Bessarabov, D., Stress tolerance assessment of dibenzyltoluene-based liquid organic hydrogen carriers. *Sustainable Energy andamp; Fuels* **2020,** *4* (9), 4662-4670.

7. Müller, K.; Aslam, R.; Fischer, A.; Stark, K.; Wasserscheid, P.; Arlt, W., Experimental assessment of the degree of hydrogen loading for the dibenzyl toluene based LOHC system. *International Journal of Hydrogen Energy* **2016,** *41* (47), 22097-22103.

8. Brückner, N.; Obesser, K.; Bösmann, A.; Teichmann, D.; Arlt, W.; Dungs, J.; Wasserscheid, P., Evaluation of Industrially applied heat-transfer fluids as liquid organic hydrogen carrier systems. *ChemSusChem* **2014,** *7* (1), 229-235.

9. Jorschick, H.; Geißelbrecht, M.; Eßl, M.; Preuster, P.; Bösmann, A.; Wasserscheid, P., Benzyltoluene/dibenzyltoluene-based mixtures as suitable liquid organic hydrogen carrier systems for low temperature applications. *International Journal of Hydrogen Energy* **2020,** *29* (45), 14897-14906.

10. Kwak, Y.; Kirk, J.; Moon, S.; Ohm, T.; Lee, Y.-J.; Jang, M.; Park, L.-H.; Ahn, C.-i.; Jeong, H.; Sohn, H.; Nam, S. W.; Yoon, C. W.; Jo, Y. S.; Kim, Y., Hydrogen production from homocyclic liquid organic hydrogen carriers (LOHCs): Benchmarking studies and energy-economic analyses. *Energy Conversion and Management* **2021,** *239*, 114124.

11. Rüde, T.; Dürr, S.; Preuster, P.; Wolf, M.; Wasserscheid, P., Benzyltoluene/perhydro benzyltoluene – Pushing the performance limits of pure hydrocarbon liquid organic hydrogen carrier (LOHC) systems. *Sustainable Energy andamp; Fuels* **2022**, *6*, 1541-1553.

12. Jorschick, H.; Dürr, S.; Preuster, P.; Bösmann, A.; Wasserscheid, P., Operational Stability of a LOHC-Based Hot Pressure Swing Reactor for Hydrogen Storage. *Energy Technology* **2019,** *7* (1), 146-152.

13. Sotoodeh, F.; Huber, B. J.; Smith, K. J., The effect of the N atom on the dehydrogenation of heterocycles used for hydrogen storage. *Applied Catalysis A: General* **2012,** *419*, 67-72.

Equation 1

$$DoH\ /\ \% = \frac{100\ \% \cdot Area_{H12BT} + 50\ \% \cdot Area_{H6BT}}{Area_{H12BT} + Area_{H6BT} + Area_{H0BT} + Area_{MeFl}}$$

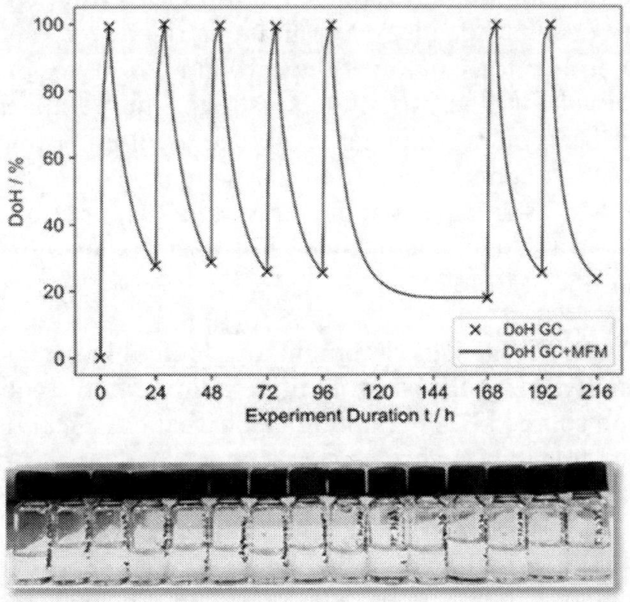

Figure 1: Top: Seven one-pot storage cycles with semi-batch hydrogenation of H0-BT and subsequent dehydrogenation of H12-BT – long-term 5th dehydrogenation over 70 h (270 g H0-BT; CI#102; 0.3wt.-% Pt/Al₂O₃ ground catalyst; n_{Pt}:$n_{H2,max}$ = 1:60000; 290 °C, $p_{H2,hy}$ = 30 bar_{H2}, $p_{H2,dehy}$ = 1.6 bar_{H2}). Bottom: 15 liquid samples taken at the start and after each hydrogenation/dehydrogenation.[11]

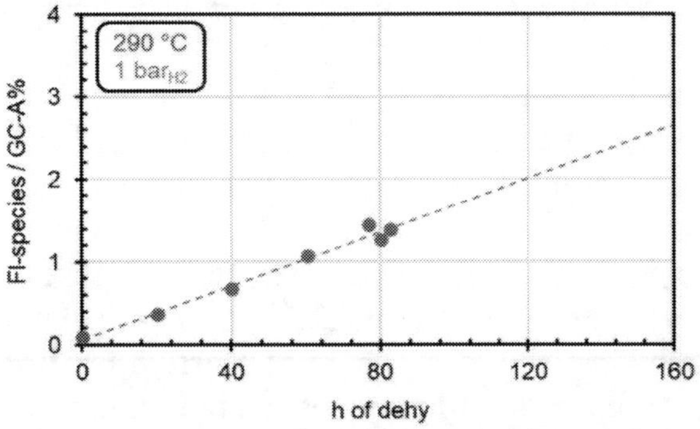

Figure 2: Fluorene-species accumulation during four dehydrogenation sequences for H18-DBT and H12-BT. Dashed lines indicate a linear trend. (270 g LOHC-; CI#102; 0.3wt.-% Pt/Al_2O_3 ground catalyst; n_{Pt}:$n_{H2,max}$ = 1:30000; 290 °C, $p_{H2,dehy}$ = 1 bar_{H2}).[11]

421a Economic Optimization of Grid-Integrated Clean Hydrogen and Ammonia Production

Matthew Palys, Prodromos Daoutidis

Using electricity generated from renewable sources to produce hydrogen via water electrolysis has been a ubiquitous topic in recent academic research, industrial development, and policy discussion. This electrified hydrogen production can lower the carbon intensity of industries which traditionally source hydrogen from methane or other fossil fuels. Perhaps more impactfully, renewable hydrogen offers vast possibilities as a carbon neutral energy storage medium or transportation fuel. Further converting this hydrogen to ammonia by combining it with air-derived nitrogen serves to increase its energy density and thus lower its storage cost in energy or transportation applications. Renewable ammonia can also be used as a sustainable nitrogen fertilizer; today, ammonia production consumes up to 50% of hydrogen produced from fossil fuels [1].

Renewable electrification of hydrogen and ammonia production has transformational potential. At present, however, this paradigm is still cost prohibitive relative to the conventional approach of using fossil fuels as the hydrogen source [2-5]. Coupling electrified chemical production with the grid is a potential approach to overcome this. As grid systems undergo a transition to include higher (and eventually dominant) fractions of renewables in their generation mix, they experience increased congestion and curtailment [6]. Time-varying chemical production as controllable load can serve to alleviate this congestion while also allowing otherwise curtailed energy to be monetized. From the chemical producer's perspective, interacting with the grid in this manner can provide low-cost or even negative-cost energy. Exploiting these potential synergies between the grid and electrified chemical producers can help drive near-term economic viability.

In this work, we optimize the economics of grid-integrated hydrogen and ammonia production. Specifically, we consider cases of (i) electrolysis-based hydrogen production and (ii) subsequent ammonia synthesis by a Minnesota transmission-scale power generation entity with natural gas, nuclear, wind, and solar generation assets. We develop a combined capacity planning and dispatch optimization model to minimize overall system costs by selecting and sizing chemical production technologies from a set of technical alternatives, for example, alkaline vs. PEM vs. solid oxide electrolysis for hydrogen production. The model simultaneously optimizes the dispatch of existing power generation assets within each of their technical (e.g., ramping, minimum up and down times) and regulatory constraints (e.g., minimum renewable generation thresholds) while also scheduling the operation of newly installed chemical production. Considering dispatch at the capacity planning stage gives rise to the above-described advantages of coupling grid operations and electrified chemical production, while the use of optimization maximizes the benefit of such integration [7].

We perform case studies for hydrogen and ammonia end products across a range of chemical production scales. We perform this analysis for present-day through 2035, considering the expected increase of renewables in the generation mix and decrease in chemical production technology costs, primarily electrolysis. For each case study, we compute the carbon intensity of the produced hydrogen or ammonia to identify trade-offs between chemical production cost and emissions in this new chemical production paradigm. Further, we consider scenarios where fossil-fueled generation assets can be retrofitted to enable hydrogen or ammonia fueling in an aim to reduce the emissions intensity of the energy supply sector. Overall, the goal of this work is to determine the viability of this grid-coupled chemical production as a path forward for near-term decarbonization of both the chemical manufacturing and energy industries.

References

[1] Lan, R., andamp; Tao, S. (2014). *Frontiers in Energy Research 2*, 35.

[2] Mallapragada, D. S., Gençer, E., Insinger, P., Keith, D. W., andamp; O'Sullivan, F. M. (2020). *Cell Reports Physical Science 1*(9), 100174.

[3] Armijo, J., andamp; Philibert, C. (2020). *International Journal of Hydrogen Energy 45*(3), 1541-1558.

[4] Nayak-Luke, R., Bañares-Alcántara, R., andamp; Wilkinson, I. (2018). *Industrial andamp; Engineering Chemistry Research 57*(43), 14607-14616.

[5] Fasihi, M., Weiss, R., Savolainen, J., andamp; Breyer, C. (2021). *Applied Energy 294*, 116170.

[6] Schermeyer, H., Vergara, C., andamp; Fichtner, W. (2018). *Energy Policy*, *112*, 427-436.

[7] Palys, M.J. andamp; Daoutidis, P. (2020). *Computers and Chemical Engineering 136*, 106785.

421b Comparative Techno-Economic and Life Cycle Analysis of Conventional Ammonia with Carbon-Capture and Low-Carbon Hydrogen-Based Ammonia

Kyuha Lee, Xinyu Liu, Pradeep Vyawahare, Pingping Sun, Amgad Elgowainy, Michael Wang

Conventional ammonia production employs fossil resources, mostly natural gas (NG), as hydrogen feedstock and accounts for more than 420 million tons of global CO_2 emissions per annum. To reduce the carbon intensity of ammonia production, alternative pathways have been studied. In this work, we investigate and compare techno-economic performances and well-to-plant-gate (WTG) GHG emissions of various ammonia production pathways, including conventional NG-based, carbon-capture, and low-carbon hydrogen-based production. Conventional NG-based ammonia refers to the ammonia produced using fossil NG through steam methane reforming and Haber-Bosch (HB) processes. Carbon-capture ammonia represents the NG-based ammonia produced while capturing process and combustion CO_2 emissions and transporting them via pipelines for CO_2 utilization. Low-carbon hydrogen-based ammonia refers to the alternative ammonia produced from carbon-free sources through water electrolysis, air separation, and HB processes using nuclear or renewable power. Most of the previous studies did not perform techno-economic and life cycle assessment for all these pathways by developing engineering process models. Also, there is a lack of studies that investigate techno-economic performances and life cycle GHG emissions of various ammonia production pathways in the United States.

In this work, we develop detailed engineering process models for each production pathway using Aspen Plus to perform techno-economic and life cycle assessment. Sensitivity analyses are conducted for uncertain parameters such as CO_2 pipeline transport distances, carbon capture tax credits, and low-carbon H_2 production costs. The results show that carbon-capture ammonia reduces WTG GHG emissions by 55–70% compared to conventional ammonia. Also, low-carbon hydrogen-based ammonia almost eliminates WTG GHG emissions because zero-carbon or near-zero carbon energy sources are employed. However, when more than \$4/kg of H_2 prices are considered for low-carbon H_2 produced using state-of-the-art electrolysis technologies, the levelized costs of low-carbon hydrogen-based ammonia are calculated to be \$0.92–\$1.06 per kg NH_3, which are about four times higher than conventional ammonia production cost. We identify that low-carbon H_2 production cost is a key cost driver for low-carbon hydrogen-based ammonia. A cheaper low-carbon H_2 production cost such as \$1/kg H_2 is needed for low-carbon hydrogen-based ammonia to be cost-competitive with conventional ammonia.

421c Techno-Economic Optimization of a Microwave-Assisted, Low-Pressure Ammonia Synthesis Process with Novel Separation Technologies

Opeyemi Ogunniyan, Chirag Mevawala, Md Emdadul Haque, Yuxin Wang, Debangsu Bhattacharyya, Jianli Hu

Ammonia is a key chemical for production of fertilizer and storing and transporting energy [1-3]. However, the traditional process for ammonia production is still the Haber-Bosch process that operates at very high temperature and pressure leading to high energy penalty and capital investment. It currently accounts for 1-2% of energy consumption, about 5% of natural gas consumption, and 1.6% of CO2 emissions worldwide[4]. The microwave (MW)-assisted ammonia synthesis process offers a viable and attractive alternative to the Haber-Bosch process, as the synthesis can occur at near atmospheric pressure and moderate temperature. MW-reactors are highly modular thus making distributed production of ammonia technically and economically feasible. Furthermore, MW reactors can be rapidly started and shutdown thus making it possible to run these systems directly with intermittent renewable energy systems. However there is currently no study on modeling, synthesis, and techno-economic optimization of the MW-assisted processes. This talk will present kinetic modeling of a MW-reactor, development of the plant-wide model, and techno-economic optimization of a modular scale ammonia synthesis process.

Our in-house lab-scale experimental data from a MW-assisted reactor is used for kinetic model development. Kinetic parameters are optimally estimated by using the experimental data. The model is validated by using the experimental data. The reactor model is then scaled up to the desired modular scale. One of the major challenges of low-pressure ammonia synthesis is the separation of ammonia from unreacted nitrogen and hydrogen at low pressure. Separation technologies such as flash separation under ambient or mildly cold conditions used in high pressure Haber-Process process are not economically viable under low pressure separation as much colder temperatures are needed for flash separation. As a result, several novel separation technologies are investigated- solid-sorbent based separation, membrane-based separation and cryogenic separation. Rigorous first-principles models of these separation technologies are developed and integrated with the MW-assisted reactor for separating and recycling unconverted reactants.

An economic model of the process is developed by using the in-house cost data for the MW-assisted reactors. Design and operating variables of the plant-wide process are optimized by an equation-oriented approach using the sequential quadratic programming approach. Since the adsorption/desorption processes used in the solid-sorbent base separation are dynamic, a surrogate model is to be used for steady-state optimization. Various economic measures such as net present value, internal rate of return, and payback period are studied by simulating a number of scenarios including the selection of separation technology, cost of reactants, and efficiency of the MW-assisted reactor.

REFERENCES

[1] E. R. Morgan, J. F. Manwell, and J. G. McGowan, "Sustainable Ammonia Production from U.S. Offshore Wind Farms: A Techno-Economic Review," ACS Sustain. Chem. Eng., vol. 5, no. 11, pp. 9554–9567, 2017, doi: 10.1021/acssuschemeng.7b02070.
[2] J. W. Erisman, M. A. Sutton, J. Galloway, Z. Klimont, and W. Winiwarter, "How a century of

ammonia synthesis changed the world," Nat. Geosci., vol. 1, no. 10, pp. 636–639, 2008, doi: 10.1038/ngeo325.
[3] C. Zamfirescu and I. Dincer, "Ammonia as a green fuel and hydrogen source for vehicular applications," Fuel Processing Technology, vol. 90, no. 5. pp. 729–737, 2009, doi: 10.1016/j.fuproc.2009.02.004.
[4] K. H. R. Rouwenhorst, P. M. Krzywda, N. E. Benes, G. Mul, and L. Lefferts, "Ammonia, 4. Green Ammonia Production," Ullmann's Encycl. Ind. Chem., pp. 1–20, 2020, doi: 10.1002/14356007.w02_w02.

421d Influence of Reductant Composition on NH_3 Synthesis from Exhaust NO Gas Using NO-CO-H_2O-H_2 Reaction

Anthony Hamzah, Hideyuki Matsumoto, Yuichi Manaka, Tetsuya Nanba

Despite being released to the atmosphere in unprecedented quantity by industrial and agricultural sectors, as well as possessing much higher global warming potentials than carbon dioxide and methane, reactive nitrogen has received relatively little attention in efforts to combat climate crisis. While NO_x in the exhaust gas is usually reduced into nitrogen that can be released without harm to the environment, this thermo-catalytic reduction process requires relatively high energy consumption. Moreover, some NO_x abatement processes (i.e. fast selective catalytic reduction) use NH_3, which is a valuable product, to reduce NO_x into N_2. On the other hand, process technology of reduction of NO_x into ammonia that could be used as a carbon-free fuel presents an attractive alternative to conventional selective catalytic reduction system and closes the nitrogen cycle.

Dissociation energy required to cleavage the N=O dual bond is much lower than, for example, triple bond in nitrogen for case of conversion of nitrogen to ammonia. Therefore, a mild operating condition for production of ammonia can be expected. Moreover, the equilibrium constant of the NO reduction is typically very favorable towards conventional synthesis of ammonia, which opens a possibility of much higher conversion and selectivity towards ammonia. Several NO_x-to-ammonia (hereinafter referred as NTA) processes have been proposed in the literature. Electrochemical reduction methods by using excess electricity from renewables have been proposed recently [1-3], however it has been deemed energetically inefficient and expensive even with assumption of declining renewable prices. On the other hand, we have reported the viability of near-complete conversion of NTA under thermocatalytic reduction process using Pt/TiO_2 catalyst at 473-523 K and atmospheric pressure [4-5]. The main reaction is shown as follows.

$NO + 2.5\,CO + 1.5H_2O \leftrightarrow NH_3 + 2.5CO_2$ (R1)

Assuming that reductant CO is supplied from the steam methane reforming (SMR) process, it is considered that produced hydrogen in the SMR could be also utilized as reductant for the NTA process.

$NO + 2.5H_2 \leftrightarrow NH_3 + H_2O$ (R2)

The use of hydrogen is expected to reduce carbon dioxide emissions in the NTA process. However, to best of our knowledge, behaviour of the NO-CO-H_2O-H_2 reaction over Pt/TiO_2 catalyst has not been fully elucidated. Thus, we investigate influence of reductant composition on ammonia synthesis using the NO-CO-H_2O-H_2 reaction by experiments and numerical analysis.

First, behavior of NO-CO-H_2O reaction on Pt/TiO_2 were investigated with fixed NO concentration of 5000 ppm, as well as CO and H_2O concentrations of 5000-20000 ppm and 8000-35000 ppm, respectively. The result showed that CO and H_2O supply above the stoichiometric level was important to guarantee conversion towards ammonia above 90%. Prior works [4-5] revealed that the reaction proceeds through formation of intermediate COOH group from CO and OH. The COOH intermediate is subsequently decomposed into CO_2 and H, of which the latter reacts with NH_x to generation of ammonia. As assuming the abovementioned reaction mechanism, we derived reaction rate model based on the Langmuir-Hinshelwood mechanism, and subsequently estimated values of parameters for the model by using the experimental data. Numerical analysis for surface coverage of intermediate on the catalyst showed that COOH coverage was constantly below 5% at relevant temperature and composition. It is estimated that an excessively high COOH coverage indicates reaction bottleneck, which limits the quantity of H to allow formation of ammonia.

Next, we investigated behavior of NO-CO-H_2O-H_2 reaction on Pt/TiO_2 by changing ratio of flow rate for H_2 under constant total flow rate of CO and H_2. NO and H_2O concentration was 5000 ppm and 7500 ppm, respectively. GHSV was set at 20000 h^{-1}. In a case when reaction temperature was 498 K, it was seen that the conversion of NO and the conversion of CO increase with increase in the flow rate ratio for H_2. It was also found that the insufficient feed of CO (below the stoichiometric proportion) yielded sub-optimal (< 50%) conversion of NO to NH_3 in either case of NO-CO-H_2O reaction and NO-CO-H_2O-H_2 reaction. Furthermore, it was shown that the selectivity of NO to NH_3 was constant at almost 100% in the presence of H_2. It was also estimated that the yield of NH_3 increased with increase in the flow rate ratio for H_2. When the reaction temperature was changed from 498 K to 473 K and 448 K, the consumption of NO declined.

It is notable that while formation of other reactive nitrogen species was investigated, production of was NO_2 not observed. On the other hand, production of N_2O, which has a higher greenhouse effect than CO_2, was confirmed. In a case when reaction temperature was 448 K, selectivity of NO to N_2O was seen to decrease with increase in the flow rate ratio for H_2. On the other hand, in a case when reaction temperature was 498 K, the selectivity of NO to N_2O was lower than the case for 448 K and increased with increase in the flow rate ratio for H_2.

Hence, it was found that in the NO-CO-H_2O reaction, the addition of H_2 has a positive effect on NH_3 production, and the effect changes depending on the reaction temperature. The above experimental results suggest that competitive adsorption of CO, H_2O and H_2 (and the resultant intermediate species) is influential towards the conversion of NO to NH_3. On the other hand, the decreasing trend of conversion towards NH_3 with increasing CO in case of fixed total flow rate of CO and H_2 implies that a correct consideration of NO-CO-H_2O composition balance is very important. Thus, we derived two reaction rate models for R1 and R2 based on the Langmuir-Hinshelwood mechanism, and subsequently estimated values of parameters for the models by using the experimental data. There was a good agreement between the calculated conversion by using estimated values of parameters and experimental data in a range of high flow rate ratio for H_2. On the other hand, in the cases of lower ratio for H_2, it was seen that calculated conversion of NO was up to ~20% lower than that of the experimental data. It is suggested more complex reaction mechanism beyond abovementioned assumptions might be at play and is a subject for further more rigorous kinetic analysis.

The presented results herein demonstrate the effectiveness of NO-CO-H_2O reaction, particularly with H_2 reductant addition, to develop more efficient NTA process. It is thought that the NTA process based on NO-CO-H_2O-H_2 reaction will become more environmentally friendly by complementarily using green hydrogen derived from renewable energy. Therefore, the future works in this study should be focused on plant-wide techno-economic analysis.

References:

[1] McEnaney, J. M., Blair, S. J., Nielander, A. C., Schwalbe, J. A., Koshy, D. M., Cargnello, M., andamp; Jaramillo, T. F. (2020). *ACS Sustainable Chemistry andamp; Engineering*, *8*(7), 2672-2681.

[2] Long, J., Chen, S., Zhang, Y., Guo, C., Fu, X., Deng, D., andamp; Xiao, J. (2020). *Angewandte Chemie International Edition*, *59*(24), 9711-9718.

[3] Daiyan R., Tran-Phu T., Kumar P., Iputera K., Tong ZZ, Leverett J., Ali Khan MH, Esmailpour AA, ..., Amal R. (2021). Energy Environ. Sci.,14, 3588-3598

[4] Kobayashi, K., Atsumi, R., Manaka, Y., Matsumoto, H., andamp; Nanba, T. (2019). *Catalysis Science andamp; Technology*, *9*(11), 2898-2905.

[5] Kobayashi, K., Ogawa, S., Matsumoto, H., Matsuda, K., andamp; Nanba, T. (2020). *Journal of the Japan Petroleum Institute*, *63*(1), 28-37.

529a Techno-Economic Evaluation of Carbon-Negative Hydrogen Production from Biomass

Mohammad Ostadi, Leslie Bromberg, Daniel R. Cohn, Emre Gençer

Hydrogen is an important energy vector which has immense potential to lower greenhouse gas (GHG) emissions in hard to abate industries. Currently H_2 is produced predominantly from fossil sources which entail associated CO_2 emissions. Biomass, on the other hand, is a natural sink for atmospheric CO_2 and if it is used for production of H_2 or other fuels, it would be almost carbon-neutral. If carbon capture is added to the process with permanent CO_2 storage, the final product would be carbon negative. This means that the process removes atmospheric CO_2. Here we evaluate a thermochemical pathway for production of H_2 from biomass (see Figure 1). It is comprised of an entrained flow gasifier (EFG), hot gas cleaning unit, water gas shift (WGS) reactor, CO_2 removal unit and pressure swing adsorption (PSA) unit. We have chosen technologies that are high in technology readiness level (TRL) and can be implemented today. The process is modeled in Aspen HYSYS® for mass and energy balances while available literature data is used to estimate associated costs of the process. The results are benchmarked against the standard steam methane reforming (SMR) with respect to economics as well as GHG emissions. Through sensitivity analysis, key factors affecting economics and emissions are highlighted. This work helps to gain insight on the viability of sustainable H_2 production from biomass. Gasification of available biomass should be considered in addition to H_2 from natural gas or water electrolysis to supply the emerging hydrogen need in industry and economy.

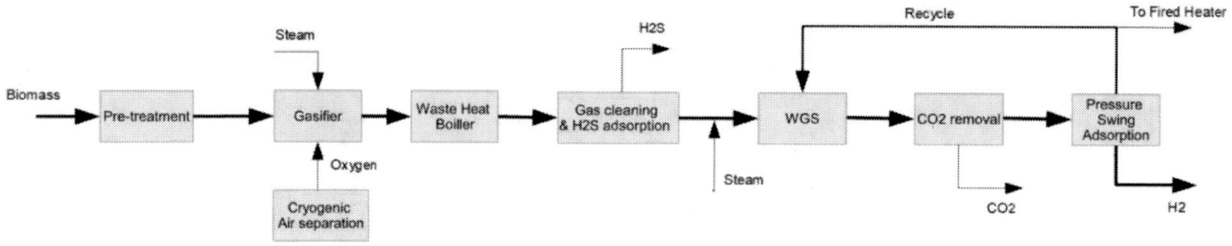

529b Hydrogen Production By Methane-Biomass Co-Pyrolysis

Haiyan Guo

Hydrogen is a promising cleanest energy carrier. Although hydrogen is a zero-emission fuel, its suitability also depends on the energy consumed and the cleanliness of the production methods. Conventional methane reforming and coal gasification process for hydrogen production are undesirable due to huge CO_2 emission and energy consumption. Biomass is considered as potential alternative renewable feedstock in the production of chemicals and fuels. As biomass partial combustion can provide heat and potential active catalysts, biochar and intermediates for methane pyrolysis without a net generation of carbon dioxide, we are interested in combining methane pyrolysis with biomass gasification. Temperature programed reaction of biomass/CH_4 copyrolysis was investigated at different heating rate to understand the synergistic effect of biomass on CH_4 decomposition. TG/MS of biomass under Ar atmosphere at different heating rate was utilized to modify the kinetic parameters form TPR and monitor intermediates releasing characteristics during biomass gasification process. BET and SEM/EDS analysis were applied to verify the function of biochar in copyrolysis biomass and methane process. Considering the energy cost, feed cost and CO_2 tax, pyrolysis, water gas shift (WGS) and partial combustion (ATR) technologies for hydrogen production were discussed. Combined the advantages them, an ideal autothermal methane-biomass partial oxidation process which is no heat addition required and makes the overall process net negative in CO_2 can serve as a cost-effective, environmentally friendly strategy to producing hydrogen.

529c Insights into Gas-Phase Methane Pyrolysis for Hydrogen Production and Carbon Capture

Patrick Lott, Manas B. Mokashi, Akash B. Shirsath, Heinz Müller, Corina Janzer, Steffen Tischer, Lubow Maier, Olaf Deutschmann

Introduction

Among technological processes in chemical industry for reducing greenhouse gas emissions, methane pyrolysis for large-scale hydrogen (H_2) production from methane (CH_4) is considered an auspicious alternative approach to water electrolysis [1, 2]. Herein, gaseous CH_4 originating from natural gas or biogas is converted into gaseous H_2 and solid carbon at high temperatures. Particularly the use of renewable biogas as feed and sustainable energy from wind and solar for setup operation would allow for achieving a net-zero emission process or even negative emissions, making methane pyrolysis a process of high relevance in both academia and industry. Despite its great potential as a CO_2-free and low-energy H_2 production route, no commercial large-scale methane pyrolysis processes have been realized so far. In addition to political aspects, primarily economic considerations and technical challenges require research and development: Efficient energy supply at high temperatures, use of materials that are suitable for high temperatures, handling of carbon deposits forming during the process and utilization of accruing carbon are only few facets that come into play in the context of methane pyrolysis. After shining a light on the boundary conditions that determine a successful development and implementation at industrial scale, the present contribution focuses on gas-phase methane pyrolysis in order to provide an overview on process parameters governing high methane conversion and hydrogen selectivity as well as on the underlying detailed chemistry.

Materials and Methods

By variation of the inlet feed gas mixture, the temperature of operation and the pressure, a systematic experimental measurement campaign was conducted using a continuous flow high-temperature, high-pressure alumina reactor (hot zone: 0.4 m) that was developed in-house. A mass spectrometer provides on-line information on the composition on the effluent product gas stream concentrations, which serve as a basis for numerical simulations with the DETCHEM^{PFR} reactor code. Assuming steady-state and isothermal conditions, numerical simulations were performed using a detailed gas-phase reaction mechanism by Appel et al. comprising a set of pyrolysis reactions of C1 and C2 species [3]. Varying the temperature between 1000 and 1600 °C, the H_2/CH_4 molar ratio in the feed between 1 and 4, the residence time between 3 and 7 s and the pressure between 1 and 8 bar provides information of the impact of the different process parameters on the pyrolysis reaction by means of both, experiments and simulations.

Results and Discussion

The experiments in our in-house developed continuous flow high-temperature, high-pressure alumina reactor coupled with a mass spectrometer revealed that gas phase methane pyrolysis takes place for temperatures above 1000 °C if using CH_4 and H_2 as feed gases. Increasing temperatures and a lower H_2/CH_4 molar ratio in the gas feed benefit CH_4 conversion and high H_2 yield (Fig. 1a). Low H_2/CH_4 feed gas ratios promote CH_4 conversion particularly between 1000 °C and 1300 °C, whereas a further temperature increase results in almost complete CH_4 conversion, irrespective of the reaction conditions. Equally important, higher residence times facilitate high H_2 yields. Among the byproducts, the numerical simulations suggest C_2H_2, C_2H_4, C_2H_6 and C_6H_6 as most important species and point to their prominent role during carbon formation in the gas phase. Although the experimental data support these findings, it becomes also clear that such intermediate gas-phase species can only survive at moderate temperatures (Fig. 1b). An increase of the pressure seems to be a suitable approach for suppressing byproduct formation, however, at the cost of lower CH_4 conversion rates.

While simulations and experiments are in good agreement at less severe conditions (i.e. low temperature and high H_2/CH_4 ratio), an almost inverse trend is observed for severe conditions. This behavior can be explained by the carbon that forms and deposits in-situ during the course of the experiment. Due to adsorption and desorption processes on the carbon layers formed, the presence of carbon significantly influences the gas-phase concentrations [4]. Future research will aim at linking the type of carbon formed and the corresponding reaction conditions, which will not only provide additional mechanistic insights, but which will also allow for optimizing the process parameters in order to produce well-defined carbon that can be utilized in industry.

References

[1] Ashik, U. P. M., Wan Daud, W. M. A., Abbas, H. F, Renewable Sustainable Energy Rev., 44 (2015) 221–256.

[2] Machhammer, O., Bode, A., Hormuth, W. Chem. Eng. Technol., 39 (2016) 1185-1193.

[3] Appel, J., Bockhorn, H., Frenklach, M., Combust. Flame, 121 (2000) 122-136.

[4] Li, A., Deutschmann, O., Chem. Eng. Sci. 62 (2007) 4976-4982.

Figure 1: a) Comparison of experiment and pure gas-phase kinetic simulations for gas-phase methane pyrolysis. b) Experimental concentration profile of gas-phase species for $H_2/CH_4 = 2$, 1 bar, $\tau = 3$ s.

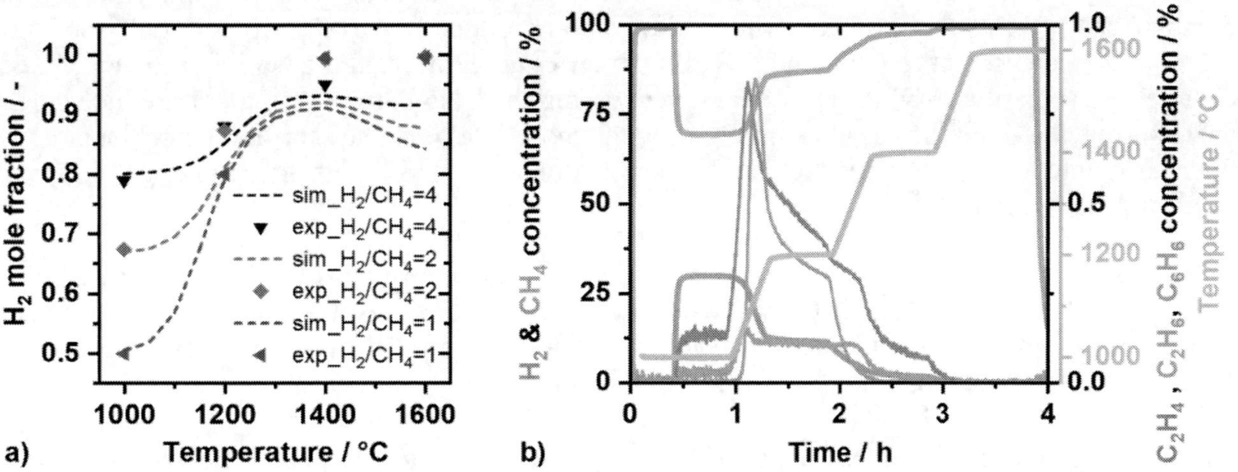

529d Methane Decarbonization for Hydrogen and Sequestered Carbon Nanofiber Co-Product

Jessica Hauck, Kent J. Warren, Linfei Li, Mija H. Hubler, Boning Wang, Andrew Broerman, Samantha Harshberger, Theodore Champ, Alan Weimer

Hydrogen is one of the most essential chemical building blocks to our energy infrastructure, with downstream applications in oil refining as well as production of ammonia, the key component in fertilizer. Today, nearly all hydrogen is commercially produced via steam methane reforming in combination with the water gas shift reaction. For every kilogram of hydrogen produced, there are roughly 9 kilograms of CO_2 released to the atmosphere.[1] With an estimated global hydrogen production rate of 75 million metric tons per year[1], over 675 million metric tons of carbon dioxide are emitted via this process. A profitable alternative to produce hydrogen while circumventing greenhouse gas (GHG) is needed. A Catalytic-Chemical Vapor Deposition (CCVD) process to produce hydrogen and carbon nanofibers (CNFs) from methane is a financially promising alternative to steam methane reforming.

Methane, the primary component of natural gas, requires substantial process conditions to overcome the strong C – H bonds. Particle Atomic Layer Deposition (ALD) enables the formation of highly dispersed metal nanocatalysts on a given support which reduce the energy required for methane conversion. Moreover, the highly dispersed catalyst, consisting of abundant metals such as iron or nickel, is essential as this process contains a "sacrificial catalyst"; the catalyst, support, and CNFs combine as to form a sequestered carbon value-added product. In partnership with both the Hubler group in the department of Civil and Environmental Engineering and the National Ready Mixed Concrete Association (NRMCA), we are investigating the CNF/catalyst product to reduce cracking and improve the durability of ultra-high purity concrete (UHPC). The sacrificial catalyst circumvents the energy intensive step of product/catalyst separation as well as avoids the challenge of rapid catalyst deactivation.

In this research, abundant transition metals are deposited onto a support via particle ALD. CNFs and hydrogen are then generated via CCVD with the particle ALD catalyst. CCVD and catalyst synthesis parameters are investigated to maximize carbon and hydrogen yield. A preliminary technoeconomic analysis is performed to assess the viability of this technology at scale.

[1]Bartlett, J., andamp; Krupnick, A. 2020. "Decarbonized Hydrogen in the US Power and Industrial Sectors: Identifying and Incentivizing Opportunities to Lower Emissions" (Report 20-25). *Resources for the Future*.

529e Microwave-Assisted Dry Reforming of Methane

Sagar Sourav, Cong Wang, Weiqing Zheng, Dionisios Vlachos

The current investigation aims at environmentally benign way of hydrogen production by utilization of renewable energy sources. The performance of ceria-based catalysts (undoped CeO_2 and Zr coped CeO_2: CZO) was investigated using microwave (MW) assisted dry reforming of methane (DRM) reaction. Steady-state catalytic data reveal that both CeO_2 and CZO catalysts are active for the DRM reaction under MW heating, but do not exhibit any noticeable activity under conventional heating by electric furnace at comparable catalyst bed temperature. Furthermore, for comparable CH_4 conversion values, the MW assisted DRM reaction consumes at least an order of magnitude lower amount of power. These observations, undoubtedly, demonstrate an efficient way of *Green Hydrogen* production through MW heating. Both Raman and UV-visible spectroscopy show that the CZO catalyst maintains a meta-stable partially reduced state during DRM reaction under MW heating and provide insights into the **active center** of the DRM reaction.

AUTHOR INDEX